U.S. Department
of Transportation
**National Highway
Traffic Safety
Administration**

DOT HS 811 020                                            August 2008

# Integrated Vehicle-Based Safety Systems

## Light-Vehicle On-Road Test Report

This document is available to the public from the National Technical Information Service, Springfield, Virginia 22161

This publication is distributed by the U.S. Department of Transportation, National Highway Traffic Safety Administration, in the interest of information exchange. The opinions, findings and conclusions expressed in this publication are those of the author(s) and not necessarily those of the Department of Transportation or the National Highway Traffic Safety Administration. The United States Government assumes no liability for its content or use thereof. If trade or manufacturers' names or products are mentioned, it is because they are considered essential to the object of the publication and should not be construed as an endorsement. The United States Government does not endorse products or manufacturers.

| | | | Form Approved |
|---|---|---|---|
| **REPORT DOCUMENTATION PAGE** | | | *OMB No. 0704-0188* |

Public reporting burden for this collection of information is estimated to average 1 hour per response, including the time for reviewing instructions, searching existing data sources, gathering and maintaining the data needed, and completing and reviewing the collection of information. Send comments regarding this burden estimate or any other aspect of this collection of information, including suggestions for reducing this burden, to Washington Headquarters Services, Directorate for Information Operations and Reports, 1215 Jefferson Davis Highway, Suite 1204, Arlington, VA 22202-4302, and to the Office of Management and Budget, Paperwork Reduction Project (0704-0188), Washington, DC 20503.

| 1. AGENCY USE ONLY (Leave blank) DOT HS 811 020 | 2. REPORT DATE August 2008 | 3. REPORT TYPE AND DATES COVERED October 2007 – February 2008 |
|---|---|---|
| 4. TITLE AND SUBTITLE Integrated Vehicle-Based Safety Systems Light-Vehicle On-Road Test Report | | 5. FUNDING NUMBERS PPA# HS-22 |
| 6. AUTHOR(S) Ryan Harrington, Andy Lam, Emily Nodine, * John J. Ference, and Wassim G. Najm | | |

| 7. PERFORMING ORGANIZATION NAME(S) AND ADDRESS(ES) U.S. Department of Transportation Research and Innovative Technology Administration Advanced Safety Technology Division John A. Volpe National Transportation Systems Center Cambridge, MA 02142 | * U.S. Department of Transportation National Highway Traffic Safety Administration Office of Vehicle Safety Research 1200 New Jersey Avenue SE Washington, DC 20590 | 8. PERFORMING ORGANIZATION REPORT NUMBER |
|---|---|---|
| 9. SPONSORING/MONITORING AGENCY NAME(S) AND ADDRESS(ES) U.S. Department of Transportation National Highway Traffic Safety Administration | | 10. SPONSORING/MONITORING AGENCY REPORT NUMBER DOT HS 811 020 |

11. SUPPLEMENTARY NOTES

| 12a. DISTRIBUTION/AVAILABILITY STATEMENT This document is available to the public through the National Technical Information Service, Springfield, Virginia 22161. | 12b. DISTRIBUTION CODE |
|---|---|

13. ABSTRACT (Maximum 200 words)

This report presents results from a series of on-road verification tests performed to determine the readiness of a prototype integrated warning system to advance to field testing, as well as to identify areas of system performance that should be improved prior to the start of the field test planned for 2009. Data was collected from tests conducted on public roads using a 2007 Honda Accord equipped with the prototype safety system. The system provides forward crash warning (FCW), lane departure warning (LDW), curve speed warning (CSW), and lane change/merge (LCM) functions, managed by an arbitration function that addresses multiple crash threats. The objectives of the on-road tests were to drive the test vehicle in an uncontrolled driving environment to measure the system's susceptibility to nuisance alerts, assess alerts in perceived crash situations, and evaluate system availability. The prototype system showed continued improvement in system performance throughout the series of tests conducted between October 2007 and February 2008. Based on positive results from the track-based verification tests conducted in February and these on-road tests, it was recommended that the light-vehicle platform proceed to field testing in Phase II. Additional adjustment of the LCM and LDW warning functions is recommended to further reduce nuisance alerts and improve system robustness.

| 14. SUBJECT TERMS Integrated vehicle-based safety systems, forward crash warning, curve speed warning, lane departure warning, lane change warning, nuisance alert, system availability, and light vehicle. | 15. NUMBER OF PAGES 36 |
|---|---|
| | 16. PRICE CODE |

| 17. SECURITY CLASSIFICATION OF REPORT Unclassified | 18. SECURITY CLASSIFICATION OF THIS PAGE Unclassified | 19. SECURITY CLASSIFICATION OF ABSTRACT Unclassified | 20. LIMITATION OF ABSTRACT |
|---|---|---|---|

NSN 7540-01-280-5500

Standard Form 298 (Rev. 2-89)
Prescribed by ANSI Std. 239-18
298-102

# METRIC/ENGLISH CONVERSION FACTORS

## ENGLISH TO METRIC

### LENGTH (APPROXIMATE)
- 1 inch (in) = 2.5 centimeters (cm)
- 1 foot (ft) = 30 centimeters (cm)
- 1 yard (yd) = 0.9 meter (m)
- 1 mile (mi) = 1.6 kilometers (km)

### AREA (APPROXIMATE)
- 1 square inch (sq in, $in^2$) = 6.5 square centimeters ($cm^2$)
- 1 square foot (sq ft, $ft^2$) = 0.09 square meter ($m^2$)
- 1 square yard (sq yd, $yd^2$) = 0.8 square meter ($m^2$)
- 1 square mile (sq mi, $mi^2$) = 2.6 square kilometers ($km^2$)
- 1 acre = 0.4 hectare (he) = 4,000 square meters ($m^2$)

### MASS - WEIGHT (APPROXIMATE)
- 1 ounce (oz) = 28 grams (gm)
- 1 pound (lb) = 0.45 kilogram (kg)
- 1 short ton = 2,000 pounds (lb) = 0.9 tonne (t)

### VOLUME (APPROXIMATE)
- 1 teaspoon (tsp) = 5 milliliters (ml)
- 1 tablespoon (tbsp) = 15 milliliters (ml)
- 1 fluid ounce (fl oz) = 30 milliliters (ml)
- 1 cup (c) = 0.24 liter (l)
- 1 pint (pt) = 0.47 liter (l)
- 1 quart (qt) = 0.96 liter (l)
- 1 gallon (gal) = 3.8 liters (l)
- 1 cubic foot (cu ft, $ft^3$) = 0.03 cubic meter ($m^3$)
- 1 cubic yard (cu yd, $yd^3$) = 0.76 cubic meter ($m^3$)

### TEMPERATURE (EXACT)
$[(x-32)(5/9)]$ °F = y °C

## METRIC TO ENGLISH

### LENGTH (APPROXIMATE)
- 1 millimeter (mm) = 0.04 inch (in)
- 1 centimeter (cm) = 0.4 inch (in)
- 1 meter (m) = 3.3 feet (ft)
- 1 meter (m) = 1.1 yards (yd)
- 1 kilometer (km) = 0.6 mile (mi)

### AREA (APPROXIMATE)
- 1 square centimeter ($cm^2$) = 0.16 square inch (sq in, $in^2$)
- 1 square meter ($m^2$) = 1.2 square yards (sq yd, $yd^2$)
- 1 square kilometer ($km^2$) = 0.4 square mile (sq mi, $mi^2$)
- 10,000 square meters ($m^2$) = 1 hectare (ha) = 2.5 acres

### MASS - WEIGHT (APPROXIMATE)
- 1 gram (gm) = 0.036 ounce (oz)
- 1 kilogram (kg) = 2.2 pounds (lb)
- 1 tonne (t) = 1,000 kilograms (kg)
- = 1.1 short tons

### VOLUME (APPROXIMATE)
- 1 milliliter (ml) = 0.03 fluid ounce (fl oz)
- 1 liter (l) = 2.1 pints (pt)
- 1 liter (l) = 1.06 quarts (qt)
- 1 liter (l) = 0.26 gallon (gal)
- 1 cubic meter ($m^3$) = 36 cubic feet (cu ft, $ft^3$)
- 1 cubic meter ($m^3$) = 1.3 cubic yards (cu yd, $yd^3$)

### TEMPERATURE (EXACT)
$[(9/5)y + 32]$ °C = x °F

## QUICK INCH - CENTIMETER LENGTH CONVERSION

## QUICK FAHRENHEIT - CELSIUS TEMPERATURE CONVERSION

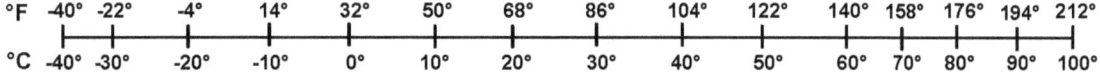

For more exact and or other conversion factors, see NIST Miscellaneous Publication 286, Units of Weights and Measures. Price $2.50 SD Catalog No. C13 10286

Updated 6/17/98

# PREFACE

The Volpe National Transportation Systems Center (Volpe Center) of the United States Department of Transportation's Research and Innovative Technology Administration is conducting an independent evaluation of integrated safety systems for motor vehicles in support of the National Highway Traffic Safety Administration (NHTSA). This research activity represents a part of the Integrated Vehicle-Based Safety Systems (IVBSS) initiative in the Intelligent Transportation Systems (ITS) program. The goal of the IVBSS program is to accelerate the deployment of integrated crash warning systems for passenger cars and heavy commercial trucks to prevent rear-end, lane change, and road departure crashes.

This report presents the results on the performance of an integrated safety system for light vehicles. Data was collected from two on-road verification tests conducted on public roads in Michigan in October 2007 and February 2008.

The authors of this report are Ryan Harrington, Andy Lam, Emily Nodine, and Wassim Najm of the Volpe Center and John J. Ference of the National Highway Traffic Safety Administration (NHTSA).

The authors acknowledge the technical contribution by Sandor Szabo of the National Institute of Standards and Technology and Al Stern of Citizant. Feedback from NHTSA reviewers is also acknowledged.

# TABLE OF CONTENTS

**1. Introduction** ................................................................................................. **10**
   1.1. System Description ................................................................................. 11
   1.2. On-Road Verification Testing .................................................................. 11
**2. Characteristics of the On-Road Verification Test** ................................ **12**
   2.1. Test Route Description ............................................................................ 12
   2.2. Road Characteristics ............................................................................... 13
   2.3. Road Type Distribution ........................................................................... 13
   2.4. Driving Maneuvers .................................................................................. 14
**3. Results of the First On-Road Test – October 2007** ............................... **14**
   3.1. Analysis of Alerts .................................................................................... 15
   3.2. Availability of Lane Departure Warning Function ................................. 18
   3.3. Conclusions From First On-Road Test .................................................... 19
**4. Results of the Second On-Road Test – February 2008** ........................ **20**
   4.1. Analysis of Alerts .................................................................................... 21
   4.2. Availability of Lane Departure Warning Function ................................. 24
   4.3. Conclusions From Second On-Road Test ............................................... 24
**5. Conclusions** ................................................................................................. **25**
**6. References** .................................................................................................. **27**
**APPENDIX A. General Guidelines for Light Vehicle On-Road Verification Tests** ............................................................................. **32**
**APPENDIX B. Definitions** .......................................................................... **35**
**APPENDIX C. Test Route Turn-by-Turn Directions** ........................... **37**

# LIST OF FIGURES

Figure 1. Map of On-Road Verification Test Route ............................................................. 13
Figure 2. Breakdown of Distance Traveled in First On-Road Test (October 2007) ........ 15
Figure 3. Breakdown of Nuisance Alert Rate by Travel Speed in First On-Road Test (October 2007) .................................................................................................................. 17
Figure 4. Breakdown of Nuisance Alert Rates in First On-Road Test (October 2007) .... 17
Figure 5. LDW Availability by Travel Speed in First On-Road Test (October 2007) ..... 19
Figure 6. Breakdown of Distance Traveled in Second On-Road Test (February 2008) ... 21
Figure 7. Nuisance Alert Rate by Travel Speed in Second On-Road Test (February 2008) .................................................................................................................................. 22
Figure 8. Breakdown of Nuisance Alert Rates in Second On-Road Test (February 2008) .................................................................................................................................. 23
Figure 9. LDW Availability by Travel Speed in Second On-Road Test (February 2008) 24
Figure 10. Breakdown of Nuisance Alert Rates for Both On-Road Tests ........................ 25
Figure 11. LDW Availability in Both On-Road Tests ....................................................... 26

# LIST OF TABLES

Table 1. Breakdown of Alerts in First On-Road Test (October 2007) ............................. 16
Table 2. Breakdown of Alerts in Second On-Road Test (February 2008) ....................... 22

## LIST OF ACRONYMS

| | |
|---|---|
| **ACAS** | Automotive Collision Avoidance System |
| **CAN** | Controller Area Network |
| **CSW** | Curve Speed Warning |
| **FCW** | Forward Crash Warning |
| **FOT** | Field Operational Test |
| **HT** | Heavy Truck |
| **IVBSS** | Integrated Vehicle-Based Safety Systems |
| **LCM** | Lane Change/Merge |
| **LDW** | Lane Departure Warning |
| **LV** | Light Vehicle |
| **NHTSA** | National Highway Traffic Safety Administration |
| **RDCW** | Road Departure Collision Warning |
| **RFA** | Request for Applications |
| **U.S. DOT** | United States Department of Transportation |

# Executive Summary

This report presents results from a series of on-road verification tests to assess the performance of an integrated safety system for light vehicles. This activity is a part of the Integrated Vehicle-Based Safety Systems (IVBSS) initiative in the Intelligent Transportation Systems (ITS) program of the U.S. Department of Transportation and addresses the prevention of rear-end, lane change, and road departure crashes. Additional information on the IVBSS program may be found on the Internet at www.its.dot.gov/ivbss/index.htm.

The goal of the IVBSS program is to accelerate the deployment of integrated crash warning systems for light vehicles[1] and heavy commercial trucks that help prevent rear-end, lane change, and road departure crashes. The prototype integrated system provides forward crash warning (FCW), lane departure warning (LDW), curve speed warning (CSW), and lane change/merge (LCM) functions and is managed by an arbitration function that addresses multiple crash threats. FCW warns drivers when they are in danger of striking the rear of the vehicle in front of them traveling in the same direction. The LDW function provides alerts to drivers when unintentionally drifting off the road edge or crossing a lane boundary. CSW alerts the driver when approaching a curve at an excessive speed. The LCM function alerts drivers when changing lanes or merging into traffic to avoid colliding with another vehicle in an adjacent lane.

The road tests used a 2007 Honda Accord equipped with the prototype warning system and was driven in an uncontrolled driving environment on public roads. Test objectives were to measure the system's susceptibility to nuisance alerts, assess alerts in perceived crash situations, and evaluate system availability over a wide range of driving conditions. Data collected during the tests was analyzed and used to evaluate system readiness for a field operational test planned for 2009 and to identify areas of system performance that could be improved prior to the start of the field test. To be ready for the field test, the prototype system must meet nuisance alert rate and LDW availability guidelines indicated in Table ES-1.

Table ES-1. IVBSS Performance Guidelines

| Performance Metric | Guidelines |
|---|---|
| Nuisance alert rate | Less than or equal to 15 nuisance alerts per 100 miles driven |
| LDW availability | 80 percent or higher on freeways<br>50 percent or higher on arterial roads<br>30 percent or higher on local roads |

---

[1] The light vehicle (LV) platform encompasses passenger cars, vans, minivans, sport utility vehicles, and light pickup trucks with gross vehicle weight ratings of 10,000 pounds or less.

On-road tests were conducted in October 2007 and February 2008. In both tests, the light vehicle prototype system's nuisance alert rate was consistently below the performance guideline of 15 nuisance alerts per 100 miles driven. However, there was a slight increase in the alert rate observed during the second test trial, attesting to variability that is characteristic of on-road tests.

Availability of the prototype vehicle's LDW function exceeded system metrics for travel speeds above 35 mph in both tests. However, the prototype's LDW availability dropped below the required performance level (22% and 16% versus the 30% performance guideline) for the lowest range of travel speeds (between 25 and 35 mph) during both test trials. These differences in LDW low-speed performance were attributed to periods of less than ideal weather during portions of the tests,[2] and to the fact that lower travel speeds are typical of rural roads, which tend to have absent or lower-quality lane markings, less than ideal lighting, and lower levels of maintenance (e.g., re-striping lane markers and removal of snow and ice during inclement weather).

It should also be noted that the prototype warning system consistently issued alerts in threatening situations that arose along each test route and correctly classified bridges, signs, and other overhead objects as non-threatening, and did not issue an alert when passing below them during both test series.

Based on continued performance improvements made throughout the test series and positive results from the track-based verification tests and these on-road tests, it was recommended that the light-vehicle platform proceed to field testing in Phase II. Adjustments to LCM and LDW alert timing were recommended to further reduce nuisance alerts and improve system robustness in all driving environments.

---

[2] During both test series, prevailing weather conditions were less than ideal for some portion of each test; the October 2007 tests had brief periods of light to moderate rain, resulting in dry, damp, and wet road surfaces, including some standing water. The February 2008 tests were performed a few days following a snowstorm in the Detroit metropolitan area; road surfaces in urban areas were clear of snow, but had areas where plowed snow covered lane markings. Rural road surfaces in some sections of the test route had heavy salt residue, which decreased contrast and recognition of lane markers; there were also isolated sections of rural roads that had packed snow and ice that completely obscured lane markers. All of these conditions contributed to a more challenging sensing environment for the LDW function and is reflected in the system performance reported.

# 1. Introduction

In November 2005, the U.S. Department of Transportation (U.S. DOT) entered into a cooperative research agreement with an industry team led by the University of Michigan Transportation Research Institute to develop and test an integrated, vehicle-based crash warning system that addresses rear-end, lane-change, and road departure crashes for light vehicles and heavy commercial trucks. The program being carried out under this agreement is known as the Integrated Vehicle-Based Safety Systems (IVBSS) program.

The goal of the IVBSS program is to assess the safety benefits and driver acceptance associated with prototype integrated crash warning systems. Preliminary analyses conducted by the National Highway Traffic Safety Administration (NHTSA) indicate that a significant number of crashes can be reduced by the widespread deployment of integrated crash warning systems that address rear-end, lateral drift, and lane change/merge crashes. Such integrated warning systems have the potential to provide comprehensive, coordinated information, from which the individual crash warning subsystems can determine the existence of a threat, and thus provide the appropriate warning to drivers.

This report presents the results of an independent assessment that examined the performance of an integrated safety system using a 2007 Honda Accord equipped with the prototype system. The test was conducted to assess the readiness of the warning system to proceed to a field operational test (FOT) that will take place in Phase II of the program, as well as to identify areas of system performance that should be improved prior to the start of the field test. This integrated warning system was designed and built for the light vehicle (LV) platform.[3] Data was collected from two on-road tests conducted on public roads in southeast Michigan under naturalistic driving conditions. Test results for the heavy truck (HT) platform on-road tests are documented in a separate report (Harrington, Lam, et al., 2008).

Several U.S. DOT staff members participated in the tests as drivers and ride-along observers. It is important to note that there may be variability in the way the system performs when being operated by other drivers due to varying driving styles and exposure to different weather, roadways, and traffic conditions. These initial tests, based on approximately 20 hours of driving and 625 vehicle miles driven, were conducted to determine if the prototype warning system was performing according to its performance guidelines; it should be noted that these results only reflect system performance for this set of drivers and do not necessarily reflect system performance for the general driving population.

To assess overall system performance and capability more thoroughly, a representative sample of drivers will be recruited to participate in a year-long field test scheduled to take place in 2009. This test will provide a larger, richer dataset from which to draw

---

[3] Light vehicles include passenger cars, vans, minivans, sport utility vehicles, and light pickup trucks with gross vehicle weight ratings of 10,000 pounds or less.

conclusions about system performance, including a significant number of vehicle miles driven. This field test, representative of 15 years of driving, will include a driver population balanced by age and gender; a wide variety of driving styles; and exposure to a broad range of weather, roadways, and traffic conditions.

## 1.2. System Description

The light-vehicle integrated system consists of the following primary crash warning functions, managed by an arbitration function that addresses multiple-crash threats (UMTRI, 2007):

- Forward crash warning (FCW) warns the driver to avoid striking the rear end of another vehicle ahead in the same lane.
- Lane departure warning (LDW) warns the driver when unintentionally drifting off the road edge or crossing a lane boundary.
- Curve speed warning (CSW) alerts the driver when approaching a curve at an excessive speed.
- Lane change/merge (LCM) alerts the driver when changing lanes or merging into traffic to avoid colliding with another vehicle in an adjacent lane, both vehicles traveling in the same direction.

All system functions are operational at vehicle speeds above 25 mph.

## 1.3. On-Road Verification Testing

The objectives of the on-road verification tests are to drive the light vehicle in an uncontrolled driving environment on public roads in order to:

- Measure the system's susceptibility to issuing nuisance alerts;[4]
- Assess alerts in perceived crash situations when they arise;
- Evaluate system availability over a wide range of driving conditions; and
- Exercise each of the four crash warning functions in order to develop a mental model and a better understanding of warning system logic.

The U.S. DOT developed the on-road test procedures and conducted the tests using Department staff as drivers and ride-along observers. These tests were devised to complement track-based verification tests utilizing an on-board data acquisition system to collect numerical and video data. Collected data was supplemented by color video recorded by an independent measurement system developed by the National Institute of Standards and Technology. The independent measurement system was installed on the test vehicle to support both track and on-road verification tests (Ference, Szabo, & Najm, 2006).

---

[4] In this document, the term "nuisance alert" is defined as a system alert that did not require immediate corrective action by the driver to avoid a collision or dangerous driving scenario. It was important to classify "nuisance" and "valid" alerts from the perspective of the driver, since ultimately the drivers' acceptance of the system relies on their perceptions of how the system works in the driving environment rather than the technical aspects of the system design.

During the on-road tests, each alert issued was classified by the driver and ride-along observers as a "nuisance" or "valid" alert using their collective subjective judgment; alerts identified in this way were later verified or reclassified through detailed, objective analysis of recorded driving data, which included target presence, and driver braking and steering behavior.

On-road verification tests were performed in October 2007 and February 2008. Tests conducted in October 2007 were repeated in February 2008 to verify that changes made to improve performance of certain system functions did not affect the performance of the overall system or other system functions. Changes implemented enhanced LCM and LDW functions and extended the FCW effective warning range for stopped vehicles. In addition, a new device to handle Controller Area Network (CAN) bus message traffic was successfully incorporated into the system design.

The characteristics of the on-road verification tests are outlined in Section 2 of this report. Results from the October 2007 and February 2008 tests are discussed in Sections 3 and 4, respectively. Section 5 provides overall conclusions of the light-vehicle on-road tests.

Guidelines for conducting the on-road tests are delineated in Appendix A. The test procedures were developed using information, experience, and prior knowledge of conditions that would elicit nuisance alerts derived from extensive experience with vehicles equipped with FCW, LCM, LDW, and CSW technologies. Previous U.S. DOT projects provided similar driving experiences from pilot and field operational tests. (Najm, Stearns, et al., 2006; Talmadge, Chu, et al., 2000; Wilson, Stearns, et al., 2007)

Appendix B defines terms used to characterize the on-road verification test procedures.

## 2. Characteristics of the On-Road Verification Test

The on-road verification test procedures consist of a structured route with fixed roadway characteristics, lighting conditions, selected maneuvers by the test vehicle, and exposure to dynamic movements of other vehicles. The selection of the driving route on public roads was based on known road characteristics and simple controllable maneuvers that can be repeated over time.

### 2.1. Test Route Description

The test route was developed using a combination of routes from previous field operational tests including the Automotive Collision Avoidance System (ACAS) and Road Departure Collision Warning (RDCW) tests (Najm, Stearns, et al., 2006; Wilson, Stearns, et al., 2007). Additional sub-routes that provided roadway characteristics needed to support IVBSS program test requirements were also included. The final test route represents a variety of roadway types in the metropolitan Detroit area that meet the general guidelines identified in Appendix B. The route was approximately 208 miles in length, and started and ended at Van Buren Township, Michigan. Figure 1 illustrates the map of the test route; turn-by-turn directions for the test route can be found in Appendix C.

## 2.2. Road Characteristics

The test route was designed to ensure that the prototype warning system would be exposed to a variety of road characteristics that are representative of normal driving for a light vehicle. The road characteristics included in the test route are listed below.

- Lane Markers: Double solid, solid, dashed, faded, and missing lane markers as well as curbs that defined lane boundaries. Numerous transitions between the different types of lane markers were also encountered.
- Number of Lanes: One and up to five lanes in the direction of travel.
- Posted Speed Limits: 25 mph to 70 mph.
- Road Geometry: Numerous curves of varying radii as well as uphill, downhill, and level grades were traversed on the route. The route included lane splits, lane merges, on and off ramps, forks, and narrow roads.
- Road Appurtenances: Jersey barriers, guardrails, mailboxes, parked cars, light poles, fences, construction barrels, and trees were present on the side of many roads. Four railroad tracks were crossed while driving the route.

## 2.3. Road Type Distribution

The test vehicle was driven in both rural and urban driving environments, with the route including 29 percent freeways, 50 percent arterial roads, and 21 percent local roads.

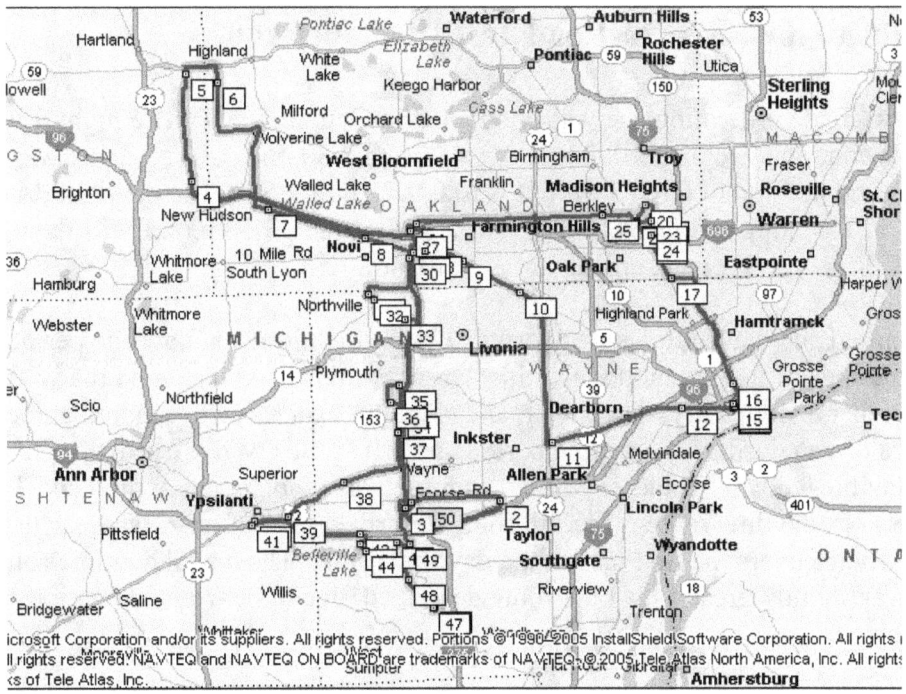

**Figure 1. Map of On-Road Verification Test Route**

### 2.4. Driving Maneuvers

Some common driving maneuvers are known to occasionally trigger nuisance alerts in crash warning systems.[5] Although nuisance alert-causing maneuvers may not actually put the vehicle in a potential crash situation, the driving scenario geometry and kinematics appear to the system like a crash scenario, thus eliciting an alert. The following is a sample of driving maneuvers that may trigger nuisance alerts:

- Passing under an overpass or overhead sign
- Approaching or negotiating a curve
- Lead vehicle turning ahead of test vehicle
- Vehicle crossing the test vehicle's path of travel
- Pulling closely behind a lead vehicle before a lane change maneuver
- Changing lanes with an adjacent vehicle two lanes over
- Pulling in front of an adjacent vehicle after a lane change
- Passing a vehicle traveling in the opposite direction with turn signal activated
- Merging and exiting the freeway
- Lanes merging or splitting

A crash warning system is expected to produce some number of nuisance alerts, but excessive nuisance alerts may cause annoyance to drivers, leading to dissatisfaction with the system. In order to address this driver acceptance issue, IVBSS performance guidelines require that the warning system shall not issue less than or equal to 15 nuisance alerts per 100 miles driven (LeBlanc, Bezzina, et al., 2008).

## 3. Results of the First On-Road Test – October 2007

The first on-road verification test was conducted in October 2007. The night drive took place on Wednesday October 10, 2007, from 5:30 to 8:30 p.m. EST. Sunset was at 6:59 p.m. EST, and the End of Civil Twilight was at 7:27 p.m. EST. The daylight drive occurred on Thursday October 11, 2007, from 8:15 a.m. to 3:30 p.m. EST. Civil Twilight began at 7:12 a.m. EST, and sunrise was at 7:40 a.m. EST.

Three-hundred and eleven miles were driven during the night and daytime periods, which included periods of light to moderate rain. These periods of rain created roadway conditions that included dry, damp, wet, and standing water. Two-hundred and eight miles were driven during the daytime period, and 103 miles were driven at night. The time-of-day breakdown for the 311-mile test route was 67 percent daytime and 33 percent nighttime. The daytime route is the full route described in Section 2, whereas the nighttime route covers the first half of the daytime route. The actual daytime route mileage was slightly greater than the route described, due to off-route deviations to refuel the test vehicle.

---

[5] Nuisance alerts refer to warnings issued in driving situations that drivers do not consider alarming and do not require an immediate corrective reaction.

The start and end times of each period allowed for driving in rush hour and non-rush hour traffic conditions, fulfilling the requirement of driving in low-, medium-, and high-traffic conditions. Figure 2 breaks down the distance traveled by travel speed bin.

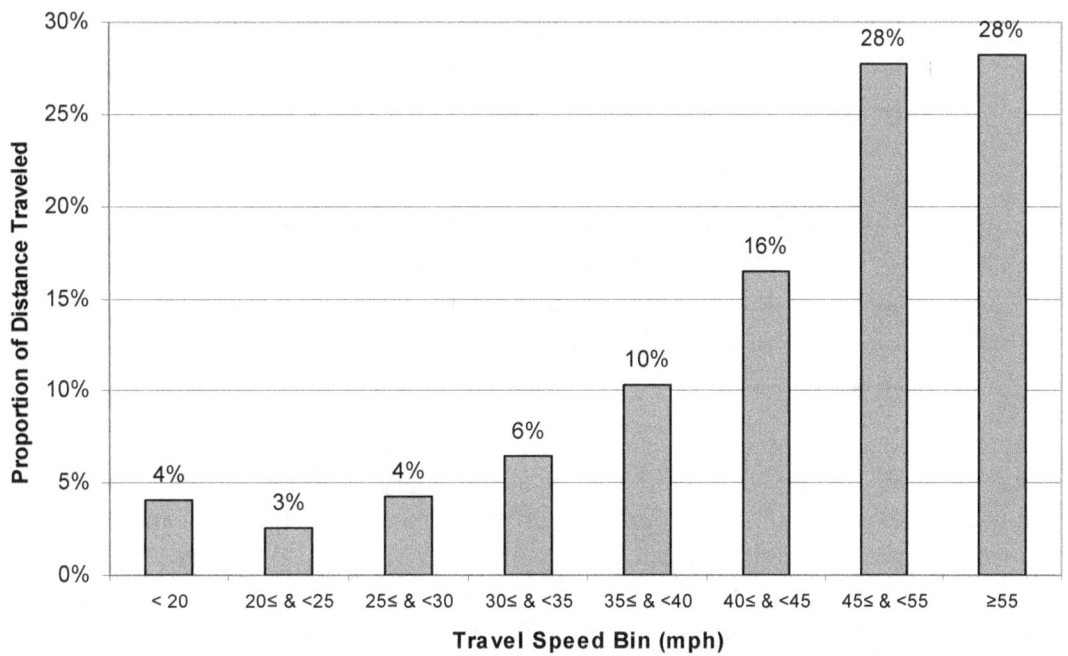

**Figure 2. Breakdown of Distance Traveled in First On-Road Test (October 2007)**

It should be noted that distribution of vehicle travel speeds in the October 2007 test was skewed toward higher travel speeds when compared to the tests performed in February 2008. This effect was largely due to the difference in prevailing weather and traffic conditions on the roadways traveled.

### 3.1. Analysis of Alerts

A total of 52 alerts were issued during the first on-road verification test – 23 alerts during the night drive and 29 alerts during the daytime drive. About 52 percent, or 27 alerts, were judged to be valid while about 48 percent or 25 alerts were considered to be nuisance alerts. During the on-road test, the driver and two ride-along observers made a subjective assessment of alert validity (valid alert or nuisance alert). A detailed and objective assessment of all alerts issued was later performed by examining the numerical and video data associated with each alert. Table 1 breaks down the valid and nuisance alerts issued for each function.

**Table 1. Breakdown of Alerts in First On-Road Test (October 2007)**

| Alert | Valid | Nuisance | Total |
|---|---|---|---|
| FCW | 4 | 4 | 8 |
| LCM-Left | 0 | 2 | 2 |
| LCM-Right | 0 | 3 | 3 |
| LDW Imminent-Left | 2 | 0 | 2 |
| LDW Imminent-Right | 1 | 4 | 5 |
| LDW Cautionary-Left | 15 | 10 | 25 |
| LDW Cautionary-Right | 3 | 0 | 3 |
| CSW | 2 | 2 | 4 |
| Total | 27 | 25 | 52 |
| % | **52%** | **48%** | **100%** |

It should be noted that an LDW alert may be a cautionary alert (i.e., drifting into an unoccupied lane) or an imminent alert (i.e., drifting to a lane occupied by another vehicle or object such as a guardrail). About 73 percent of LCM and LDW alerts were to the left direction, with the majority of these alerts being LDW cautionary alerts. Nuisance alerts accounted for 64 percent of all LCM and LDW alerts to the right direction, compared to 41 percent of all LCM and LDW alerts to the left.

Figure 3 illustrates the system-level nuisance alert rate per 100 miles by travel speed bin. While the majority of nuisance alerts were issued at speeds above 25 mph, the speed at which all warning functions become active, a few alerts were issued below 25 mph due to system communications delays.

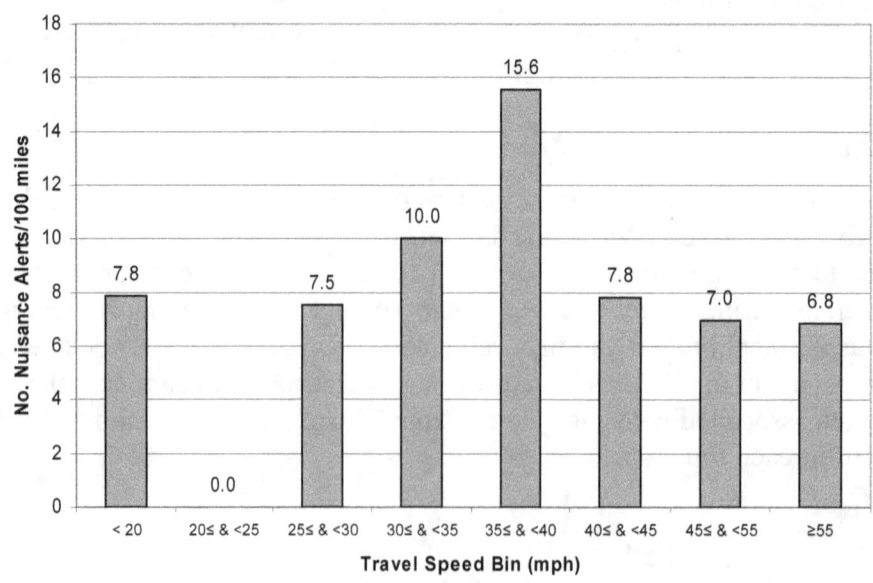

**Figure 3. Breakdown of Nuisance Alert Rate by Travel Speed in First On-Road Test (October 2007)**

Figure 4 illustrates the system-level nuisance alert rate per 100 miles and for each warning function. Overall, the nuisance alert rate was 8.0 nuisance alerts per 100 miles driven. All alerts were issued during the course of normal driving and were not triggered by intentional maneuvers. The red line in Figure 4 indicates the nuisance alert performance guideline of 15 alerts per 100 miles (LeBlanc, Bezzina, et al., 2008).

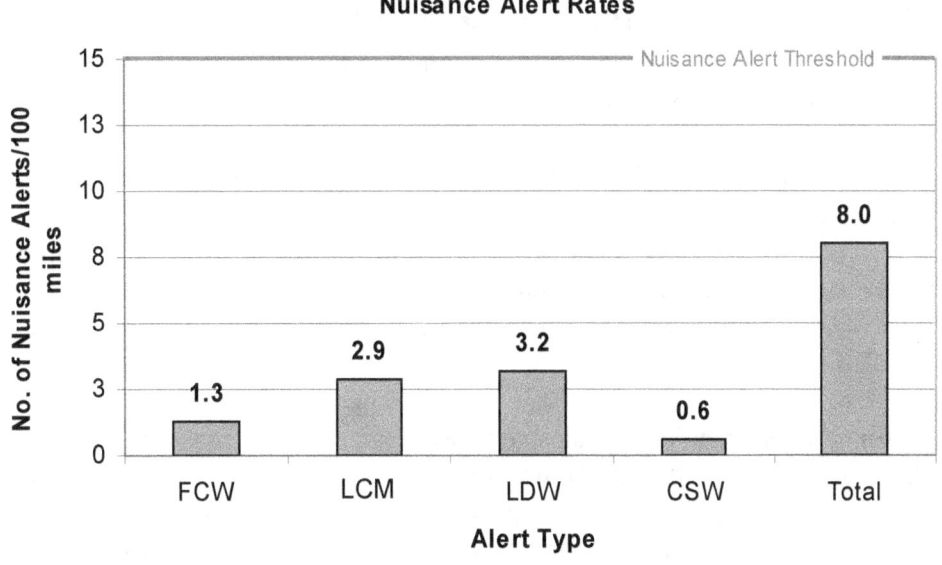

**Figure 4. Breakdown of Nuisance Alert Rates in First On-Road Test (October 2007)**

FCW nuisance alerts were attributed to vehicles ahead of the test vehicle that:

- Cut across the test vehicle's path;
- Turned right at an intersection; or
- Decelerated or started from a stopped position at a stop sign.

LCM nuisance alerts were issued:

- During a passing maneuver (i.e., passing a vehicle, then returning to the same travel lane in front of the vehicle just passed);
- In situations where right lane markers were absent, and an unknown roadside object was detected and no apparent drift of the test vehicle was perceived by the driver and ride-along observers; or
- In one isolated case where the vehicle speed was below the 25 mph warning threshold.

The following contributed to LDW nuisance alerts:

- Poor lane tracking in rain due to reduced contrast with road surfaces
- Lane splitting from one to two lanes

CSW nuisance alerts were issued while negotiating moderate curves on exit ramps. The alert timing was judged by the driver and ride-along observers as somewhat conservative since the vehicle entered the curve at a moderate but safe speed, with adequate time and distance to slow down. It is recognized that conservative alert timing could be a conscious design decision on the part of the system developer.

## 3.2. Availability of Lane Departure Warning Function

The IVBSS Request for Applications (RFA) specified the following LDW availability performance guidelines for each road type (NHTSA 2005):

- Freeways (speed limit above 55 mph) – greater than 80 percent of distance traveled on freeways
- Arterials (speed limit between 35 and 55 mph) – greater than 50 percent of distance traveled on arterial roads
- Local (speed limit between 25 and 35 mph) – greater than 30 percent of distance traveled on local roads

The LDW function is considered available when it is tracking a roadway's left and right lane markers. This enables the function to issue crash alerts that are associated with lateral lane drifting events. As seen in Figure 5, the LDW function exceeded the availability performance guidelines for freeways and arterial roads. However, LDW system performance was slightly below the guideline for local roads, where it was available 22 percent of the distance traveled, versus the 30-percent guideline. This minor performance shortfall could be attributed to periods of less than ideal weather conditions that prevailed during the test; there were brief periods of light to moderate rain that created damp and wet roads with isolated sections of standing water, thereby producing more challenging conditions for the LDW function to correctly identify lane markings with high confidence and reliability.

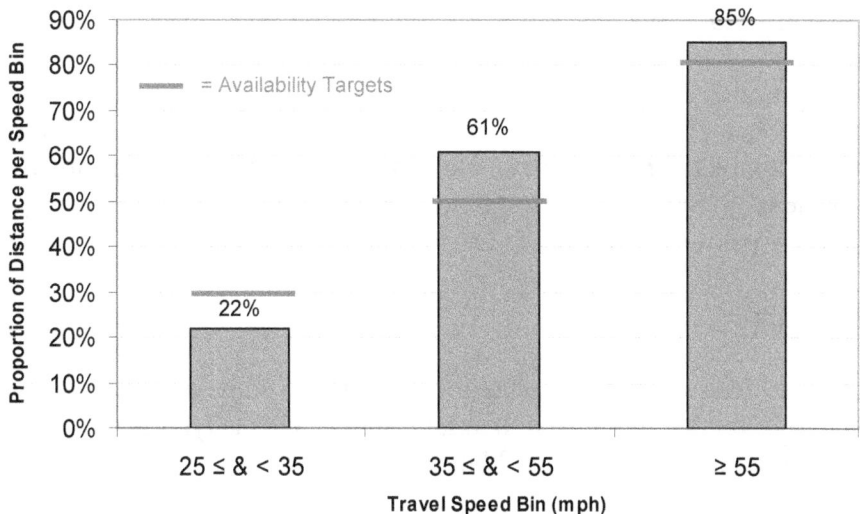

**Figure 5. LDW Availability by Travel Speed in First On-Road Test (October 2007)**

## 3.3. Conclusions From First On-Road Test

Results from this first on-road verification test demonstrated the initial capability of the light-vehicle prototype safety system and identified some sources of nuisance alerts as well. The initial nuisance alert rate observed was quite promising; at 8 nuisance alerts per 100 miles driven, it was below the system guideline of 15 or fewer nuisance alerts per 100 miles driven. In addition, the prototype system showed performance improvements over similar systems tested in past U.S. DOT-sponsored field tests. For example, the FCW function exhibited improved rejection of out-of-path targets, as well as consistent rejection of bridges, signs, and other overhead objects when compared to similar field-tested systems. LDW system availability for vehicle speeds above 35 mph exceeded system guidelines. However, some improvement in the LDW function should be considered so that system availability for the most challenging speed range, between 25 and 35 mph, is addressed.

To reduce the number of nuisance alerts and provide more robust system performance, changes to following areas should be considered:

### 1. FCW – Adjustments to alert timing

Some alerts for lead vehicle decelerating or starting in traffic ahead were judged to be issued conservatively (i.e., when the driver felt there was sufficient time to react to the lead vehicle with only moderate braking). If possible, additional logic or tuning to delay warnings for these scenarios should be considered. It is also recommended that discrimination of lead vehicles turning ahead at intersections be investigated to further reduce nuisance alerts. These are FCW issues that are commonly associated with the current generation of FCW products and prototypes due to the challenge of real-time sensing and prediction of vehicle motion in dynamic situations.

### 2. LCM – Reduction of nuisance alerts during passing maneuvers

After passing another vehicle in an adjacent lane, the system seems to hold onto that target too long. Nuisance alerts were issued when safe and typical lane changes were completed in front a vehicle that was recently passed during a safely executed maneuver. System tuning should be performed to allow lane changes in front of recently passed vehicles without issuing nuisance alerts.

### 3. LDW – Better lane tracking

Erroneous identification of lane marker locations led to some nuisance alerts especially during rainy conditions and at lane splits. This is typically an issue with the current generation of vision-based lane tracking systems.

### 4. System service message

There were some unexpected system shutdowns experienced due to receiving this message during testing. This was attributed to bus traffic issues with messages at different transmission rates. NOTE: The root cause of this problem was identified and subsequently addressed prior to conduct of the second on-road verification test.

## 4. Results of the Second On-Road Test – February 2008

Following system changes to enhance the LDW function and improve system-level performance, a second on-road verification test was conducted in February 2008. This test series was performed to verify improvements made and to measure overall system on-road performance. The night drive took place on Tuesday February 19, 2008, from 5:25 p.m. to 7:45 p.m. EST under mostly cloudy skies. Sunset was at 6:09 p.m. EST, and the End of Civil Twilight was at 6:38 p.m. EST. The daylight drive was conducted on Wednesday February 20, 2008, from 8:20 a.m. to 1:45 p.m. EST under skies ranging from mostly cloudy to partly sunny with gusty winds. Civil Twilight began at 6:54 a.m. EST, and sunrise was at 7:22 a.m. EST.

A total of 306 miles were driven during the nighttime and daytime periods on roads that had been snow covered several days before testing. During testing, the roads were clear of snow, except for some instances where there was residual snow on the roadside covering lane markings. Many of the local and arterial roads had road salt residue, though the lane markers could still be seen through it; however, the contrast between the road surface and the lane markings was reduced.

A total of 203 miles were driven in the daytime drive and 103 miles were driven at night. The time-of-day breakdown for the 306-mile test route was 67 percent daytime and 33 percent nighttime. Figure 6 provides a breakdown of the distance traveled by travel speed bin.

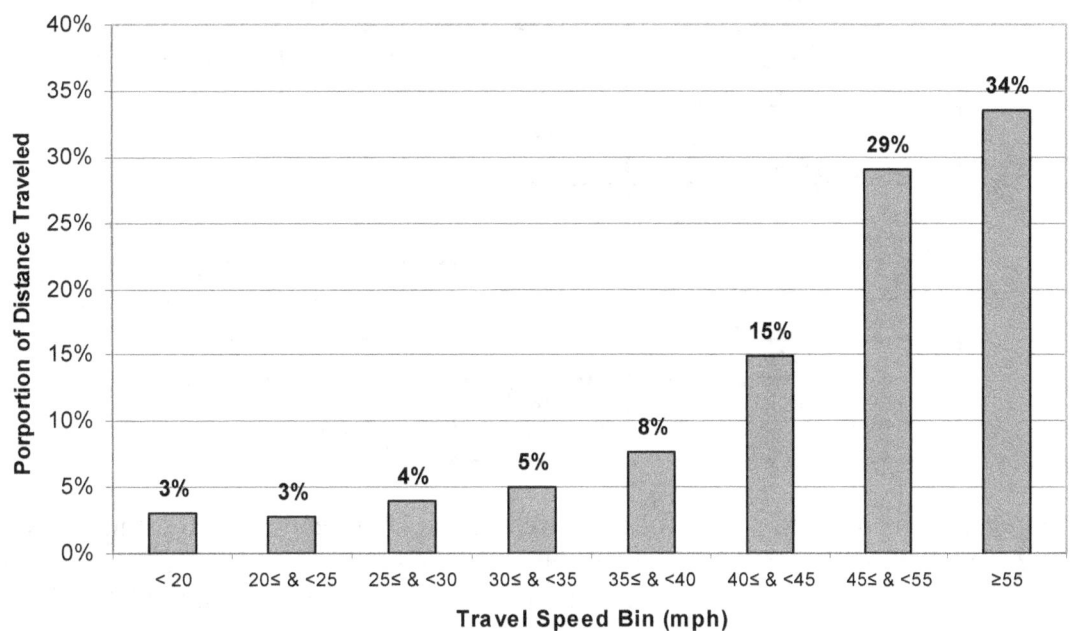

**Figure 6. Breakdown of Distance Traveled in Second On-Road Test (February 2008)**

### 4.1. Analysis of Alerts

A total of 49 alerts were issued during the test: 23 alerts during the nighttime drive and 26 alerts during the daytime drive. About 29 percent, or 14 alerts, were judged to be valid, while about 71 percent, or 35 alerts, were considered to be nuisance alerts. During the on-road test, the driver and two ride-along observers made a subjective assessment of alert validity (valid alert or nuisance alert). A detailed and objective assessment of all alerts issued was later performed by examining the numerical and video data associated with each alert. Table 2 provides a breakdown of the valid and nuisance alerts issued by each function.

Table 2. Breakdown of Alerts in Second On-Road Test (February 2008)

| Alert | VALID | NUISANCE | TOTAL |
|---|---|---|---|
| FCW | 0 | 1 | 1 |
| LCM-Left | 0 | 5 | 5 |
| LCM-Right | 1 | 4 | 5 |
| LDW Imminent-Left | 1 | 3 | 4 |
| LDW Imminent-Right | 0 | 1 | 1 |
| LDW Cautionary-Left | 12 | 8 | 20 |
| LDW Cautionary-Right | 0 | 9 | 9 |
| CSW | 0 | 4 | 4 |
| Totals | 14 | 35 | 49 |
| % | 29% | 71% | 100% |

About 66 percent of LCM and LDW alerts were to the left direction, with the majority of these alerts being LDW cautionary alerts. Nuisance alerts accounted for 93 percent of all LCM and LDW alerts to the right direction, compared to 55 percent of all LCM and LDW alerts to the left.

Figure 7 illustrates the system-level nuisance alert rate per 100 miles by travel speed bin. While most of the nuisance alerts were issued at speeds above 25 mph, the speed at which all warning functions become active, a few alerts were issued below 25 mph due to system communications delays.

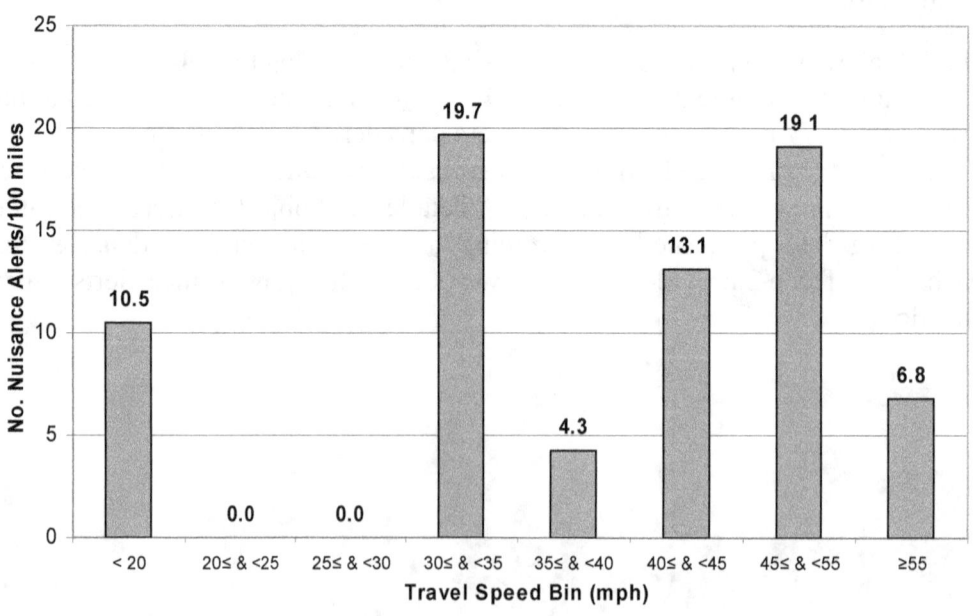

Figure 7. Nuisance Alert Rate by Travel Speed in Second On-Road Test (February 2008)

Figure 8 illustrates the system-level nuisance alert rate per 100 miles for each warning function. When compared to the October 2007 test, the nuisance alert rate increased slightly (8 nuisance alerts per 100 miles versus 11.4 nuisance alerts per 100 miles), but was still below the system guideline.

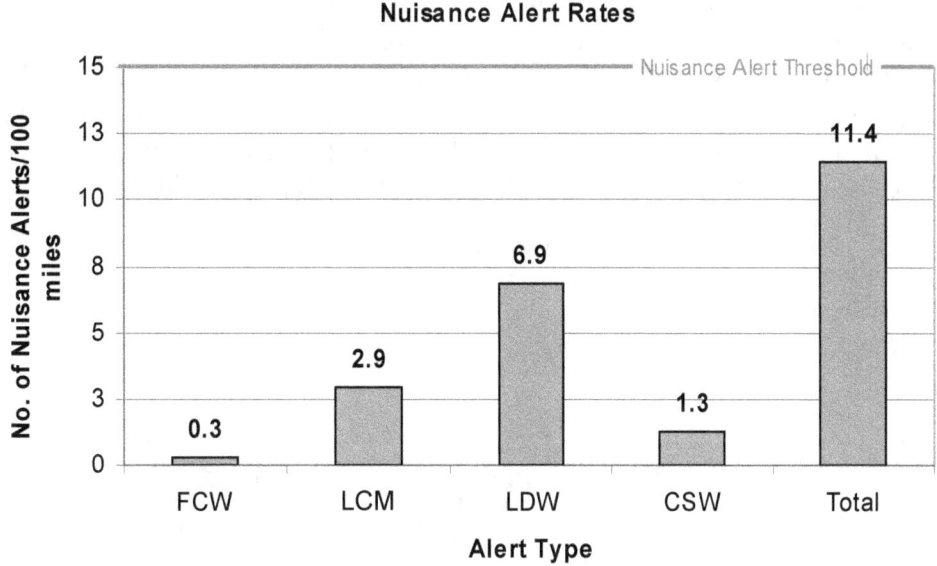

**Figure 8. Breakdown of Nuisance Alert Rates in Second On-Road Test (February 2008)**

One FCW nuisance alert was attributed to tracking an unidentified stopped object.

The following contributed to LCM nuisance alerts:

- Spurious alerts caused by improper CAN bus traffic filtering
- During a passing maneuver (i.e., passing a vehicle, then returning to the same lane of travel in front of the vehicle just passed)

The following contributed to LDW nuisance alerts:

- Spurious alerts caused by improper CAN bus traffic filtering
- Tracking visual features other than the true lane edges (e.g., tracking a high contrast road feature, such as black tar to fill cracks in the road surface, within the lane markings, or problems associated with glare from oncoming traffic temporarily "blinding" LDW sensors)
- Issuing imminent alerts when moving toward a clear shoulder or lane, while the vehicle is within the lane markers

As in the October 2007 test, CSW alerts were issued while negotiating moderate curves on exit ramps. Once again, the alerts issued were judged as nuisances because the driver and ride-along observers felt that the vehicle was traveling at a moderate but safe speed, with adequate time and distance to slow down.

## 4.2. Availability of Lane Departure Warning Function

As seen in Figure 9, the LDW function exceeded the availability performance guideline for both freeway and arterial roads. However, system performance fell below the guidelines for lower speeds (between 25 and 35 mph), where it was available 16 percent of the distance traveled, versus the 30 percent guideline. This reduction in LDW availability could be attributed to the presence of a heavy salt residue on many road surfaces, which reduced the contrast between lane markings and the road, making it more difficult for the LDW function to correctly identify lane markings with high confidence.

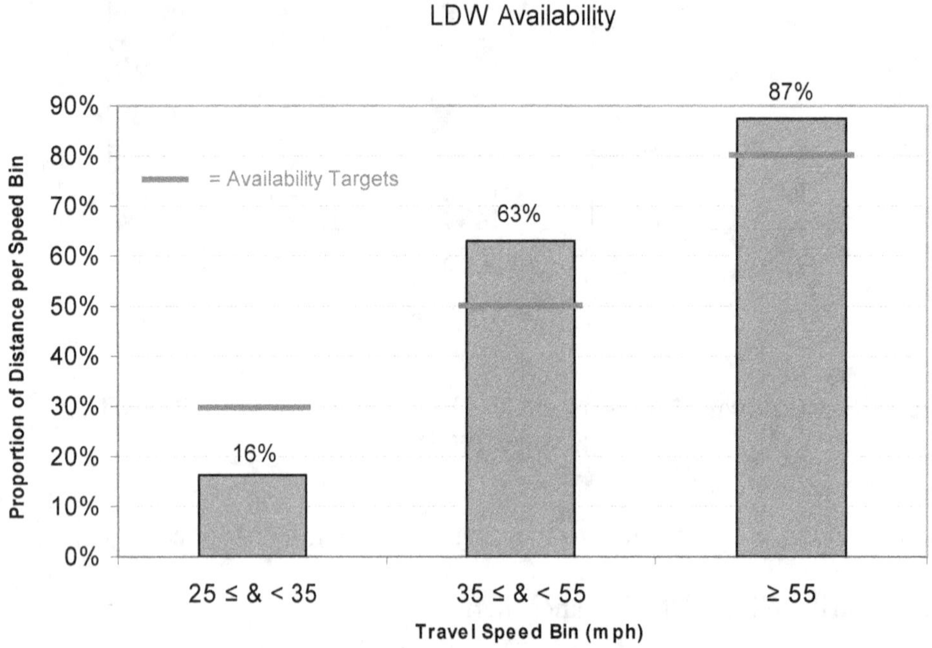

**Figure 9. LDW Availability by Travel Speed in Second On-Road Test (February 2008)**

## 4.3. Conclusions From Second On-Road Test

Light-vehicle system performance for the second on-road test was consistent with results observed during earlier tests; the nuisance alert rate measured was still below the system guideline, but increased due to slight increases in LDW and CSW nuisance alerts, and excellent rejection of out-of-path targets and bridges, signs, and other overhead objects continued. LDW system performance was also comparable; system availability that exceeded the guideline for vehicle speeds above 35 mph was maintained and LDW availability at speeds between 25 and 35 mph still was still below the system metric. Examination of the following scenarios observed during the second on-road tests could result in further reduction of the prototype's susceptibility to nuisance alerts and more robust performance before field testing in Phase II:

1. Spurious LCM and LDW alerts linked to improper CAN bus traffic filtering

2. LCM alerts issued during passing maneuvers

3. Isolated cases of LDW alerts issued when the test vehicle was traveling within a lane adjacent to a clear shoulder

## 5. Conclusions

The light-vehicle prototype warning system showed improvement and consistent performance during the on-road verification test series. In both tests, the prototype's nuisance alert rate met performance guidelines. Data collected also demonstrated the prototype system's ability to consistently issue alerts each time a threatening situation arose, and to reject out-of-path targets, bridges, signs and other overhead objects as threats when encountering them on the road. By the end of the test series, 208 bridges and 20 overhead signs were encountered, resulting in no nuisance alerts issued. Figure 10 shows the system-level and nuisance alert rate for each warning function for both on-road tests.

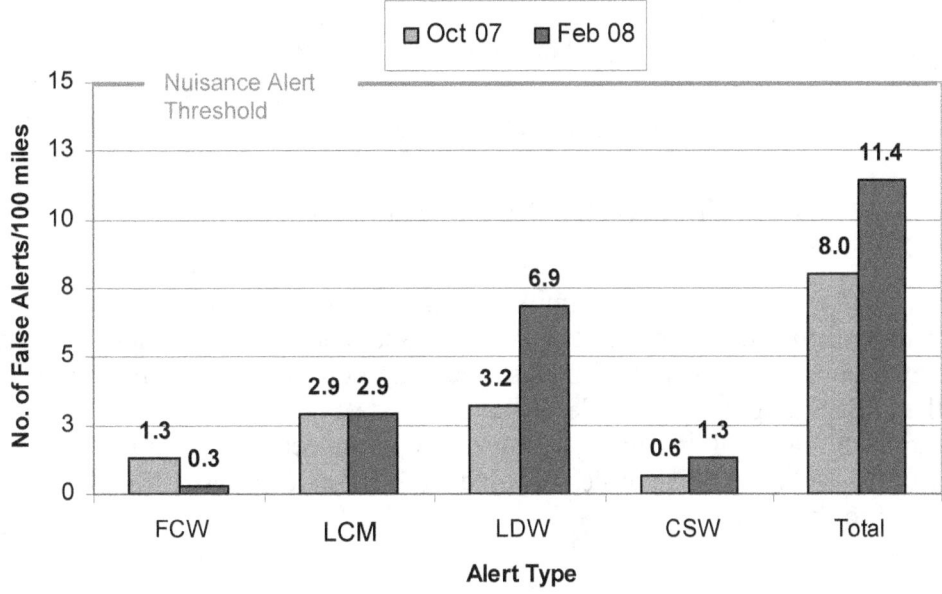

Figure 10. Breakdown of Nuisance Alert Rates for Both On-Road Tests

As mentioned earlier, the LDW system availability function was identical for both tests; for travel speeds above 35 mph performance guidelines were exceeded, while performance for speeds between 25 and 35 mph fell below the system guideline (22% and 16%, respectively, versus the 30% guideline) in both cases. The shortfall in lower speed LDW performance could be attributed to periods of less than ideal weather during portions of the tests [6] and to the fact that lower travel speeds are typical of rural roads, which tend to have absent or lower-quality lane markings, less than ideal lighting, and lower levels of maintenance (e.g., re-striping lane markers and removal of snow and ice during inclement weather). Figure 11 presents the LDW system availability results for both on-road tests.

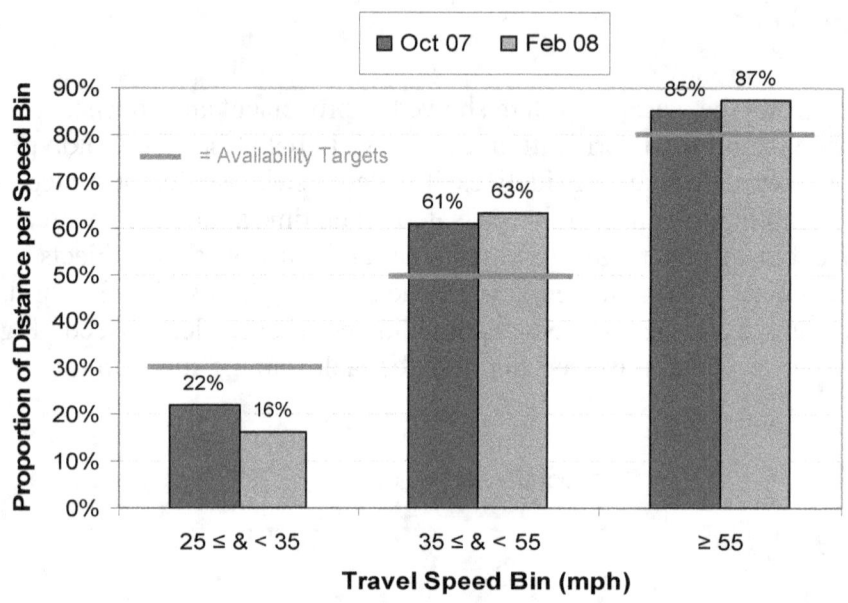

**Figure 11. LDW Availability in Both On-Road Tests**

While these on-road tests provided a preliminary look at the light vehicle prototype warning system performance, a more comprehensive assessment will be conducted by an independent evaluation of a 12-month field test planned to take place in 2009 during Phase II of the IVBSS program. The field test will include a larger and more varied driver population and range of driving styles (over 100 volunteer drivers will participate), system exposure of 200,000 vehicle miles traveled, and a broader range of weather, roadway and traffic conditions.

---

[6] During both test series, prevailing weather conditions were less than ideal for some portion of each test; the October 2007 tests had brief periods of light to moderate rain, resulting in dry, damp, and wet road surfaces, including some standing water. The February 2008 tests were performed a few days following a snowstorm in the Detroit metropolitan area; road surfaces in urban areas were clear of snow, but had areas where plowed snow covered lane markings; rural road surfaces in some sections of the test route had heavy salt residue which decreased contrast and recognition of lane markers; there were also isolated sections of rural roads that had packed snow and ice that completely obscured lane markers. All of these conditions contributed to a more challenging sensing environment for the LDW function and is reflected in the system performance reported.

# 6. References

1. Ference, J.J., Szabo, S., and Najm, W.G. (2006). Performance Evaluation of Integrated Vehicle-Based Safety Systems. *Proceedings of the Performance Metrics for Intelligent Systems (PerMIS) Workshop.* Gaithersburg, MD: National Institute of Standards and Technology.
2. Harrington, R.J., Lam, A.H., Nodine, E.E., and Ference, J.J., Najm, W.G. (2008). Integrated Vehicle-Based Safety Systems Light Vehicle On-Road Verification Test Report (DOT HS 811 020). Washington, DC: U.S. Department of Transportation, National Highway Traffic Safety Administration.
3. LeBlanc, D., Bezzina, D., Tiernan, T., Freeman, K., Gabel, M. and Pomerleau, D.: System Performance Guidelines for a Prototype Integrated Vehicle-Based Safety System (IVBSS) – Light Vehicle Platform." University of Michigan Transportation Research Institute technical report UMTRI-2008-20, March 2008.
4. Najm, W.G., Stearns, M.D., Howarth, H., Koopmann, J., and Hitz, J. (2006). Evaluation of an Automotive Rear-End Collision Avoidance System (DOT HS 810 569). Washington, DC: U.S. Department of Transportation, National Highway Traffic Safety Administration.
5. National Highway Traffic Safety Administration. (2005). Discretionary Cooperative Agreement for Integrated Vehicle-Based Safety Systems (IVBSS), Request for Applications. Washington, DC: U.S. Department of Transportation.
6. Talmadge, S., Chu, R., Eberhard, C., Jordan, K., and Moffa, P. (2000). Development of Performance Specifications for Collision Avoidance Systems for Lane Change Crashes (DOT HS 809 414). Washington, DC: U.S. Department of Transportation, National Highway Traffic Safety Administration.
7. Transportation Research Board. (2000). Highway Capacity Manual 2000.
8. University of Michigan Transportation Research Institute. (2007). Integrated Vehicle-Based Safety Systems First Annual Report (DOT HS 810 842). Washington, DC: U.S. Department of Transportation, National Highway Traffic Safety Administration.
9. Wilson, B.H., Stearns, M.D., Koopmann, J., and Yang, C.Y.D. (2007). Evaluation of a Road-Departure Crash Warning System (DOT HS 810 854). Washington, DC: U.S. Department of Transportation, National Highway Traffic Safety Administration.

# APPENDIX A. General Guidelines for Light-Vehicle On-Road Verification Tests

The following guidelines were used to develop the on-road verification test route. They were developed using information and experience obtained from the Automotive Collision Avoidance System and Roadway Departure Collision Warning field operational tests.

## A.1. Driving Environment

### A.1.1. Road Type and Land Use

The test route shall include freeway, arterial and local roadway types located in urban and rural areas that represent typical light vehicle driving patterns. The route length shall be a minimum of 200 miles and be distributed as follows:

- Road Type:
    - 25-35 percent freeway (speed limit 45-75 mph)
    - 45-55 percent arterial (speed limit 35-50 mph)
    - 20-30 percent local (speed limit 15-35 mph)
- Land Use:
    - 50-60 percent urban; and
    - 40-50 percent rural

### A.1.2. Light Conditions

Outside light conditions shall include daylight, darkness, and dusk and artificial lighting, such as streetlights, that represent typical conditions encountered by the vehicle. The lighting conditions on the test route shall contain 65 to 75 percent daylight and 25 to 35 percent nighttime driving.

The daytime route shall include a two-hour period in the early morning and a two-hour period in the late afternoon. Early morning starts two hours after dawn and late afternoon ends two hours before twilight. Night driving shall be conducted two hours after twilight. Dawn, dusk, and twilight times are available from the U.S. Naval Observatory Web site (http://aa.usno.navy.mil/data/).

### A.1.3. Traffic Conditions

The test vehicle should encounter low, moderate, and heavy traffic conditions, corresponding to specific service levels defined by the Highway Capacity Manual 2000 as follows (Transportation Research Board, 2000):

- Low traffic: Service levels A and B
- Moderate traffic: Service levels C and D
- Heavy traffic: Service levels E and F

The test route shall be planned in order to be exposed to these three levels of traffic conditions.

A.1.4. Weather Conditions

The test shall be conducted on days when clear weather, without precipitation, predominates; clear skies, without or with a few scattered clouds, are also preferred.

**A.2. Driving Scenarios**

Driving scenarios, which shall exercise each subsystem warning function, shall be executed on the test route as described below.

A.2.1. Exposure Scenarios

The vehicle is traveling on a straight road or on a curve, without making any maneuvers, and is exposed to the following roadway features:

- Fixed Features:
    - Curves: small (radius of curvature less than 500 m); medium (radius of curvature between 500 and 1000 m); and large (radius of curvature over 1000 m)
    - Profile: level, downhill, and uphill (greater than 1% grade)
    - Side objects: Jersey barrier, guardrail, sign, mailboxes, pole, tree, bridge support or abutment, parked car, etc., within 2 m of the travel lane
    - Overhead objects: bridge, sign, etc.
    - Surface objects: metal covers, train tracks, etc.
    - Lane markers: good markers on both sides, markers on one side, faded markers
    - Road layout: narrow street, ramp, fork, lane split, lane merge, etc.
- Dynamic Features:
    - Other vehicles turning, changing lanes, cutting across the light-vehicle, etc.

A.2.2. Maneuvers by Test Vehicle:

The test driver shall safely initiate a variety of driving maneuvers, such as lane changes, turns, merges, passing, etc.

**A.3. Driver Guidelines**

A.3.1. Driver

The test vehicle shall be driven by an "independent driver" who is not part of the industry project team, nor related to team members or suppliers of system components. An observer shall accompany the driver to provide navigation instructions and take real-time notes of alerts issued by the system. Detailed, objective analysis of these alerts shall be performed later using data collected by an on-board independent measurement system and a data acquisition system.

## A.3.2. Driving Behavior

The driver:
- Shall obey all posted speed limits and drive in a normal, naturalistic manner;
- May perform maneuvers that are considered part of normal driving (e.g., change lanes in heavy traffic, closely follow a lead vehicle at greater than two-second headway, etc.);
- Shall not attempt to induce warning conditions (e.g., accelerate into lead vehicle), unless scripted in the on-road test procedures; and
- Shall conduct all maneuvers, naturalistic or scripted, in a safe manner without posing any risk to the test vehicle, its passengers, other vehicles or pedestrians.

# APPENDIX B. Definitions

## B.1. Alert Descriptions

### B.1.1 Valid Alert

Valid alerts refer to warnings issued for driving situations that most drivers would consider threatening and would require an immediate corrective action to avoid a collision or dangerous situation.

### B.1.2 Nuisance Alert

Nuisance alerts refer to warnings issued for driving situations that most drivers would not consider threatening and would not require an immediate corrective action by the driver. There are three types of nuisance alerts, as follows:

- System-related nuisance alerts: caused by internal system noise or processing artifacts, when there is no object or threat present.
- In-path nuisance alerts: caused by other vehicles that are in the path of the equipped vehicle, but are at a distance or moving at a speed that most drivers do not perceive as threatening. For example, forward crash warnings are issued for lead vehicles turning right or left at intersections. Some of these alerts could be issued as part of a conservative system design, but some drivers may perceive the alerts as unnecessary.
- Out-of-path nuisance alerts: caused by vehicles and objects that are not in the equipped vehicle's path.

## B.2. Road Types

The following is NAVTEQ's categorization of roadway functional classes that were used for the conduct of the on-road tests:

Level 1. Roads with very few, if any speed changes, typically controlled access, and provide high-volume, maximum speed movement between and through major metropolitan areas.
Level 2. Roads with very few, if any, speed changes, and those that provide high-volume, high-speed traffic movement. Typically used to channel traffic to (and from) Level 1 roads.
Level 3. Roads that interconnect Level 2 roads and provide a high volume of traffic movement at a lower level of mobility than Level 2 roads.
Level 4. Roads that provide for a high volume of traffic movement at moderate speeds between neighborhoods.
Level 5. All other roads.

Levels 1 and 2 are mostly freeways; Level 3 is considered an arterial road, while Levels 4 and 5 refer to local roads.

**B.3. Land Use**

Land use classifies populated areas as either urban or rural. An urban area is one where streets are located within a developed locale (i.e., an area that has increased density of human-created structures compared to areas surrounding it). Urban areas may be cities or towns, but the definition is not commonly extended to rural settlements such as villages or hamlets.

A rural area (also referred to as "the country" or "the countryside") is a settled place outside towns and cities. Such areas are distinct from more intensively settled urban and suburban areas, and also from unsettled lands such as the outback, American Old West or wilderness. Inhabitants live in villages, hamlets, on farms and in other isolated houses.

# APPENDIX C. Turn-by-Turn Directions of Light-Vehicle Test Route

| Mile | Instruction | For | Toward |
|---|---|---|---|
| 0.0 | **Depart Visteon Way, Belleville, MI 48111 on Visteon Way (North)** | 65 yds | |
| 0.1 | Turn RIGHT (East) onto Ecorse Rd | 5.4 mi | |
| 5.4 | Turn RIGHT (South) onto Middlebelt Rd | 0.2 mi | |
| **5.6** | **At 7507 Middlebelt Rd, Romulus, MI 48174, stay on Middlebelt Rd (South)** | **0.3 mi** | |
| 6.0 | Keep RIGHT onto Ramp | 0.6 mi | I-94 / Chicago |
| 6.5 | Take Ramp (LEFT) onto I-94 | 3.5 mi | I-94 / Chicago |
| 10.1 | At exit 194, turn RIGHT onto Ramp | 0.2 mi | I-275 / Toledo / Flint |
| 10.3 | Take Ramp (RIGHT) onto I-275 | 2.5 mi | I-275 / Flint |
| **12.7** | **At 20, stay on I-275 (North)** | **15.6 mi** | |
| 28.3 | At exit 165, take Ramp (LEFT) onto I-96 | 15.2 mi | I-96 / Lansing |
| 43.5 | At exit 151, turn RIGHT onto Ramp | 0.2 mi | Kensington Rd |
| 43.7 | Keep LEFT to stay on Ramp | 65 yds | |
| 43.8 | Turn LEFT (North) onto Kensington Rd | 0.7 mi | |
| **44.5** | **At 4880 Kensington Rd, Milford, MI 48380, stay on Kensington Rd (North)** | **2.0 mi** | |
| 46.5 | Turn LEFT to stay on Kensington Rd | 1.6 mi | |
| 48.1 | Road name changes to (S) Pleasant Valley Rd | 2.6 mi | |
| 50.7 | Keep STRAIGHT onto (N) Pleasant Valley Rd [Pleasant Valley Rd] | 0.6 mi | |
| **51.3** | **At 1563 N Pleasant Valley Rd, Hartland, MI 48353, stay on N Pleasant Valley Rd [Pleasant Valley Rd] (North)** | **0.4 mi** | |
| 51.8 | Turn RIGHT (East) onto M-59 [W Highland Rd] | 1.9 mi | |
| 53.7 | Turn RIGHT (South) onto S Hickory Ridge Rd | 1.1 mi | |
| **54.7** | **At S Hickory Ridge Rd, stay on S Hickory Ridge Rd (South)** | **0.9 mi** | |
| 55.7 | Road name changes to N Hickory Ridge Trail | 2.0 mi | |
| 57.7 | Turn LEFT (East) onto General Motors Rd [GM Rd] | 2.5 mi | |
| 60.2 | Turn RIGHT (South) onto (S) Milford Rd | 5.0 mi | |
| 65.2 | Turn LEFT (East) onto Grand River Ave | 0.8 mi | |
| **66.0** | **At 55453 Grand River Ave, New Hudson, MI 48165, stay on Grand River Ave (East)** | **6.1 mi** | |
| **72.1** | **At 44408 Grand River Ave, Novi, MI 48375, stay on Grand River Ave (East)** | **6.2 mi** | |
| **78.3** | **At Grand River Ave, stay on Grand River Ave (East)** | **1.6 mi** | |
| 79.9 | Merge onto M-5 [Grand River Ave] | 2.4 mi | |
| **82.3** | **At near Redford, stay on M-5 [Grand River Ave] (East)** | **1.5 mi** | |
| 83.7 | Turn RIGHT (South) onto US-24 [Telegraph Rd] | 8.7 mi | |
| 92.4 | Bear RIGHT (South-West) onto Ramp | 0.1 mi | US-12 / Michigan Ave |
| 92.6 | Turn RIGHT (East) onto US-12 [Michigan Ave] | 0.3 mi | |
| **92.9** | **At near Dearborn Heights, stay on US-12 [Michigan Ave] (East)** | **8.3 mi** | |
| **101.1** | **At near Detroit, stay on US-12 [Michigan Ave] (East)** | **3.2 mi** | |
| **104.3** | **At near Detroit, stay on US-12 [Michigan Ave] (East)** | **164 yds** | |
| 104.4 | Bear RIGHT (South) onto M-1 [Woodward Ave] | 109 yds | |
| 104.5 | Take Local road(s) (LEFT) onto M-1 [Woodward Ave] | 0.2 mi | |
| **104.7** | **At near Detroit, stay on M-1 [Woodward** | **0.2 mi** | |

| | | | |
|---|---|---|---|
| | Ave] (North) | | |
| **104.9** | **At near Detroit, stay on M-1 [Woodward Ave] (North-West)** | **98 yds** | |
| 105.0 | Turn RIGHT (North-East) onto Witherell St | 142 yds | |
| 105.0 | Turn RIGHT (East) onto Madison St | 0.4 mi | |
| 105.4 | Take Ramp (RIGHT) onto I-75 [Chrysler Fwy] | 1.1 mi | I-75 / North Fisher Fwy |
| **106.5** | **At near Detroit, stay on I-75 [Chrysler Fwy] (North)** | **6.6 mi** | |
| 113.1 | At exit 59, take Ramp (RIGHT) onto Oakland St | 0.3 mi | M-102 / 8 Mile Rd |
| 113.4 | Bear LEFT (North) onto Local road(s) | 0.3 mi | M-102 / 8 Mile Rd |
| 113.7 | Merge onto E 8 Mile Rd | 0.2 mi | |
| 113.9 | Merge onto M-102 [E 8 Mile Rd] | 1.0 mi | |
| 114.9 | Keep STRAIGHT onto E 8 Mile Rd | 174 yds | M-1 / Woodward Ave |
| **115.0** | **At E 8 Mile Rd, Ferndale, MI 48220, stay on E 8 Mile Rd (West)** | **0.1 mi** | |
| 115.1 | Turn RIGHT (North) onto Woodward Ave | 0.2 mi | |
| 115.4 | Merge onto M-1 [Woodward Ave] | 2.7 mi | |
| 118.0 | Turn RIGHT (East) onto W Lincoln Ave | 0.3 mi | |
| **118.3** | **At 335 W Lincoln Ave, Royal Oak, MI 48067, stay on W Lincoln Ave (East)** | **0.1 mi** | |
| 118.5 | Turn LEFT (North) onto (S) Main St | 1.3 mi | |
| **119.8** | **At near Royal Oak, stay on N Main St (North)** | **0.6 mi** | |
| 120.4 | Turn LEFT (West) onto Vinsetta Blvd | 0.4 mi | |
| **120.8** | **At near Berkley, stay on Vinsetta Blvd (West)** | **1.2 mi** | |
| **122.0** | **At near Huntington Woods, stay on Vinsetta Blvd (South-West)** | **21 yds** | |
| 122.0 | Turn LEFT (East) onto Lawndale Dr | 65 yds | |
| 122.0 | Turn RIGHT (South) onto Iroquois Blvd | 153 yds | |
| **122.1** | **At near Huntington Woods, stay on Iroquois Blvd (South)** | **54 yds** | |
| 122.1 | Turn LEFT (East) onto Catalpa Dr | 0.8 mi | |
| **123.0** | **At 978 N Washington Ave, Royal Oak, MI 48067, turn RIGHT (South) onto (N) Washington Ave** | **0.9 mi** | |
| **123.8** | **At near Royal Oak, turn RIGHT (West) onto W 6th St** | **0.4 mi** | |
| 124.2 | Turn RIGHT (North-West) onto M-1 [Woodward Ave] | 1.7 mi | |
| 125.9 | Keep LEFT onto Local road(s) | 32 yds | |
| 125.9 | Bear LEFT (South-East) onto M-1 [Woodward Ave] | 0.1 mi | |
| 126.0 | Turn RIGHT (West) onto (W) 12 Mile Rd | 1.7 mi | |
| **127.7** | **At 15673 W 12 Mile Rd, Southfield, MI 48076, stay on W 12 Mile Rd (West)** | **11.3 mi** | |
| 139.0 | Keep LEFT onto Country Club Dr | 0.1 mi | |
| **139.1** | **At 39001 Sunrise Dr, Farmington, MI 48331, stay on Country Club Dr (South)** | **0.6 mi** | |
| 139.7 | Turn LEFT (South) onto Haggerty Rd | 0.3 mi | |
| 140.0 | Turn LEFT (East) onto Hills Tech Dr | 0.2 mi | |
| **140.2** | **At 38900 Hills Tech Dr, Farmington, MI 48331, stay on Hills Tech Dr (East)** | **0.8 mi** | |
| 141.0 | Turn RIGHT (South) onto Halsted Rd | 1.5 mi | |
| **142.5** | **At 24466 Halsted Rd, Farmington, MI 48335, stay on Halsted Rd (South)** | **0.2 mi** | |
| 142.7 | Turn RIGHT (West) onto Grand River Ave | 0.3 mi | |
| 143.0 | Turn RIGHT to stay on Grand River Ave | 0.4 mi | |
| **143.4** | **At 38936 Grand River Ave, Farmington, MI 48335, stay on Grand River Ave (West)** | **0.3 mi** | |
| 143.7 | Turn LEFT (South) onto Haggerty Rd | 0.5 mi | |
| **144.2** | **At 23670 Haggerty Rd, Farmington, MI 48335, stay on Haggerty Rd (South)** | **1.8 mi** | |
| 146.0 | Turn RIGHT (West) onto (E) 8 Mile Rd [Base Line Rd] | 2.5 mi | |

| | | | |
|---|---|---|---|
| 148.5 | Turn LEFT (South) onto N Center St | 0.4 mi | |
| **148.9** | **At 370 N Center St, Northville TWP, MI 48167, stay on N Center St (South)** | **0.2 mi** | |
| 149.1 | Turn LEFT (East) onto (E) Main St [Northville Rd] | 0.6 mi | |
| **149.7** | **At S Main St, Northville TWP, MI 48167, stay on S Main St [Northville Rd] (South)** | **0.3 mi** | |
| 149.9 | Road name changes to Northville Rd | 0.9 mi | |
| 150.9 | Turn LEFT (East) onto 6 Mile Rd | 1.5 mi | |
| **152.4** | **At 40104 6 Mile Rd, Northville TWP, MI 48167, stay on 6 Mile Rd (East)** | **0.6 mi** | |
| 153.0 | Take Ramp (RIGHT) onto I-275 [I-96] | 5.9 mi | I-275 / I-96 |
| **158.9** | **At 25, turn off onto Ramp** | **0.4 mi** | **M-153 / Ford Rd / Westland / Garden City** |
| 159.4 | Turn RIGHT (West) onto M-153 [Ford Rd] | 0.2 mi | |
| 159.5 | Turn RIGHT (North) onto N Haggerty Rd | 2.1 mi | |
| 161.6 | Turn LEFT (West) onto Joy Rd | 0.1 mi | |
| **161.7** | **At 41135 Joy Rd, Canton, MI 48187, stay on Joy Rd (West)** | **0.6 mi** | |
| 162.3 | Turn LEFT (South) onto N Lilley Rd | 1.0 mi | |
| 163.3 | Turn LEFT (East) onto Warren Rd | 0.1 mi | |
| **163.5** | **At near Plymouth, stay on Warren Rd (East)** | **0.7 mi** | |
| 164.1 | Turn RIGHT (South) onto (N) Haggerty Rd | 1.7 mi | |
| **165.8** | **At near Canton, stay on N Haggerty Rd (South)** | **2.1 mi** | |
| 167.9 | Turn RIGHT (West) onto US-12 [Michigan Ave] | 3.5 mi | |
| **171.4** | **At US-12, stay on US-12 [Michigan Ave] (West)** | **4.0 mi** | |
| **175.4** | **At US-12, Ypsilanti, MI 48198, stay on US-12 (South-West)** | **120 yds** | |
| 175.5 | Take Ramp onto I-94 [US-12] | 1.8 mi | I-94 |
| 177.3 | At exit 183, turn RIGHT onto Ramp | 0.3 mi | Huron St / Ypsilanti |
| **177.6** | **At US-12 Bus, Ypsilanti, MI 48197, take Local road(s) (RIGHT) onto US-12 Bus [S Hamilton St]** | **0.2 mi** | **Huron St South / Whittaker Rd** |
| 177.7 | Keep LEFT onto Ramp | 0.1 mi | I-94 / US-12 / Detroit |
| 177.9 | Keep LEFT to stay on Ramp | 10 yds | |
| **177.9** | **At near Ypsilanti, stay on Ramp (North)** | **87 yds** | **I-94 / US-12 / Detroit** |
| 178.0 | Merge onto I-94 [US-12] | 6.8 mi | |
| **184.8** | **At 190, turn RIGHT onto Ramp** | **0.4 mi** | **Belleville Rd / Belleville** |
| 185.2 | Turn RIGHT (South) onto Belleville Rd | 0.8 mi | |
| **186.0** | **At Belleville Rd, Belleville, MI 48111, stay on Belleville Rd (South)** | **65 yds** | |
| 186.0 | Road name changes to Main St | 0.5 mi | |
| **186.6** | **At 29 Main St, Belleville, MI 48111, stay on Main St (South-East)** | **32 yds** | |
| 186.6 | Turn LEFT (North-East) onto (W) Huron River Dr | 2.0 mi | |
| **188.5** | **At 41827 E Huron River Dr, Belleville, MI 48111, stay on E Huron River Dr (East)** | **0.6 mi** | |
| 189.1 | Turn RIGHT (South) onto Haggerty Rd | 2.4 mi | |
| 191.5 | Road name changes to Savage Rd | 2.0 mi | |
| 193.5 | Turn LEFT (North) onto Gentz Rd, then immediately turn RIGHT (East) onto S Metro Pkwy | 0.2 mi | |
| 193.7 | Turn RIGHT (South) onto Waltz Rd | 0.1 mi | |
| **193.8** | **At near New Boston, stay on Waltz Rd (South)** | **0.3 mi** | |
| 194.1 | Turn RIGHT (West) onto Judd Rd | 0.2 mi | |
| 194.4 | Turn RIGHT (North) onto Gentz Rd | 0.1 mi | |
| **194.5** | **At near New Boston, stay on Gentz Rd (North)** | **0.2 mi** | |
| 194.7 | Turn LEFT (West) onto Savage Rd | 2.0 mi | |
| 196.7 | Road name changes to Haggerty Rd | 0.2 mi | |
| **196.9** | **At near New Boston, stay on Haggerty** | **2.1 mi** | |

|       | **Rd (North)** |         |
|-------|----------------|---------|
| 199.1 | Turn RIGHT (East) onto E Huron River Dr | 0.3 mi |
| **199.4** | **At E Huron River Dr, Belleville, MI 48111, stay on E Huron River Dr (North-East)** | **87 yds** |
| 199.4 | Turn LEFT (North-West) onto Haggerty Rd | 2.9 mi |
| 202.3 | Turn RIGHT (East) onto Ecorse Rd | 0.7 mi |
| 203.0 | Turn RIGHT (South) onto Visteon Way | 65 yds |
| **203.0** | **Arrive Visteon Way, Belleville, MI 48111** | |

DOT HS 811 020
August 2008

U.S. Department of Transportation
**National Highway Traffic Safety Administration**

www.ingramcontent.com/pod-product-compliance
Lightning Source LLC
Chambersburg PA
CBHW081803170526
45167CB00008B/3309